BORN OF HEAT AND PRESSURE
Mountains and Metamorphic Rocks

Patricia L. Barnes-Svarney

—an Earth Processes book—

ENSLOW PUBLISHERS, INC.
Bloy St. & Ramsey Ave.
Box 777
Hillside, N.J. 07205
U.S.A.

P.O. Box 38
Aldershot
Hants GU12 6BP
U.K.

Copyright © 1991 by Patricia L. Barnes-Svarney

All rights reserved.

No part of this book may be reproduced by any means without the written permission of the publisher.

Library of Congress Cataloging-in-Publication Data

Barnes-Svarney, Patricia L.
    Born of heat and pressure: mountains and metamorphic rocks / Patricia L. Barnes-Svarney.
        p. cm. — (An Earth processes book)
    Summary: Describes the formation of different types of mountains and metamorphic rock (rocks that change form in the intense heat and pressure of the Earth's interior) and discusses their impact on their environment.
       ISBN 0-89490-276-8
    1. Rocks, Metamorphic—Juvenile literature. 2. Orogeny—Juvenile literature. [1. Rocks, Metamorphic. 2. Mountains. 3. Geology.] I. Title. II. Series.
QE475.A2B365 1991   89-25856
552'.4—dc20   CIP
                                                                              AC

Printed in the United States of America

10 9 8 7 6 5 4 3 2 1

Illustration Credits:
American Heritage Center, University of Wyoming, pp. 43, 54, 57; Arizona Meteor Crater, p. 47; Patricia L. Barnes-Svarney, pp. 7, 8, 12, 16, 17, 19, 22, 27, 28, 30, 31, 35, 39; A. Keith, U.S. Geological Survey, p. 26; Marie Morisawa, p. 34; National Aeronautics and Space Administration, pp. 48, 49; J.R. Stacy, U.S. Geological Survey, p. 52; David W. Tuttle, pp. 37, 38; Wyoming Travel Commission, pp. 4, 9.

Cover Photograph:
Wyoming Travel Commission

Contents

1 The Mystery of Metamorphic Rocks
  and Mountains 5
2 Clues in Hot Rocks 14
3 Press and Stress 24
4 Gneiss and Other Rocks 33
5 Mountains Around the World
  (And Elsewhere) 41
6 Living with Metamorphic Rocks 50
  Glossary 59
  Further Reading 62
  Index 63

The Wind River Range in Wyoming is one of the most beautiful mountain ranges in the United States.

1

The Mystery of Metamorphic Rocks and Mountains

From space, the earth looks like a peaceful and serene planet. The blue-green oceans cover most of the globe, interrupted by an occasional brown continent. A closer look at the earth's landmasses reveals their most prominent features, from the flat plains of the central United States to the rugged mountains of the European Alps.

Scientists know that deep under the surface of this tranquil planet lies a dynamic and fascinating world, where the moving interior is dominated by heat and pressure. These hot layers and the tremendous pressures inside the earth have played an important part in the formation of metamorphic rocks and mountains.

It was a long time before anyone understood the interior of the earth. In ancient times, some humans lived in caves. But these openings to the earth were very shallow. No one knew what was deep inside the earth, but there was much speculation. Legends grew of gnomes and dwarfs who mined the tunnels under the earth's surface. Other stories described the earth as hollow, with a great, intelligent race of beings living inside!

Others believed that the earth's interior was a harsh world known as the "underworld" or a place filled with heat and horror. The Bible speaks of the eternal fires of an underworld called Hades. Dante, the famous Italian poet, wrote of touring the underworld, where darkness and terror were everywhere. Many ancient Greek writings describe huge subterranean caverns where hot sulfuric winds would blow and the ground would tremble.

Many ancient peoples were particularly fearful of volcanoes or places where hot rock (magma) from deep inside the earth reaches the surface. For example, in 1500 B.C., the volcano Thira (also called Santorini) in Greece spit out more hot rocks and gas than ever recorded. Not far away, the Italian cities of Pompeii and Herculaneum were destroyed in 79 A.D. after an eruption of the volcano Vesuvius.

In 1678, Athanasius Kircher, a German scholar, wrote that the earth's interior was a raging fire. He thought that hot springs and geysers, or the boiling, bubbling surface water seen around active volcanic areas, were caused by water that had seeped in from the oceans and had come in contact with the earth's internal fires. He also believed erupting volcanoes were really tongues of fire that had reached the earth's surface.

Today geologists, or scientists who study the earth, know a great deal about what really happens at the earth's interior. They know how mountains grow and how most surface features evolve. They have many theories, or ideas, about the motions of the earth's interior. Most importantly, they know how rocks and minerals form on the earth's surface and deep in the interior.

A Look at Rocks and Minerals

What is a mineral? It is a naturally occurring solid composed of one or more chemical elements. Elements are small inorganic particles, such as oxygen, iron, silica, or potassium. There are more than 100 known elements. For example, the mineral quartz is made up of the chemical elements silica and oxygen.

There are more than 1,300 minerals in the earth's crust. Each has certain characteristics depending on the elements found in the material. Some minerals are very soft, while others are very hard. Some minerals are heavy, and others are very light. Each mineral breaks in a certain way, such as the scallop-shaped breaks in obsidian glass and the way mica splits into flat sheets.

One of the most important characteristics of a mineral is its hardness. If a rock is made up of many hard minerals, the rock will not be eroded easily by exposure to such elements as wind and water. Whereas, a rock made of softer minerals will easily break down.

In order to keep track of a mineral's hardness, Friedrich Mohs developed a special scale in 1812. Mohs' scale of hardness has ten minerals arranged in their order of hardness. Diamonds top the scale as the hardest mineral. Talc is at the bottom of the list; it can easily be scratched with a fingernail. Few minerals have a hardness exactly matching any of the minerals on the scale; most lie somewhere between two of the minerals on the chart.

What are rocks? They are natural masses of material that make up the earth's crust. Each rock is formed under certain specific conditions and is composed of one or more minerals. Granite, for example, is a rock made up of the minerals quartz, mica, hornblende, and feldspar. Marble

MOHS' SCALE OF HARDNESS

1. Talc	6. Orthoclase
2. Gypsum	7. Quartz
3. Calcite	8. Topaz
4. Fluorite	9. Corundum
5. Apatite	10. Diamond

The minerals on Mohs' scale of hardness.

is mainly made of the mineral calcite, with traces of other minerals.

The study of rocks and minerals is very important to geologists who want to know more about the earth. From the composition of the minerals, the way they lie within the rock, and their size, geologists can often discover the history of a rock. Some rocks, such as the natural glass obsidian, form by the rapid cooling of magma (molten rock). Others, such as conglomerates and breccias, form from deposits of eroded rock. And still others, such as gypsum, form when water evaporates, leaving a mineral behind—much like the salt crystals found in a pan when saltwater evaporates.

Geologists divide the earth's rocks into three families: igneous, sedimentary, and metamorphic. Each rock family is formed under different conditions and has certain characteristics. Each falls some-

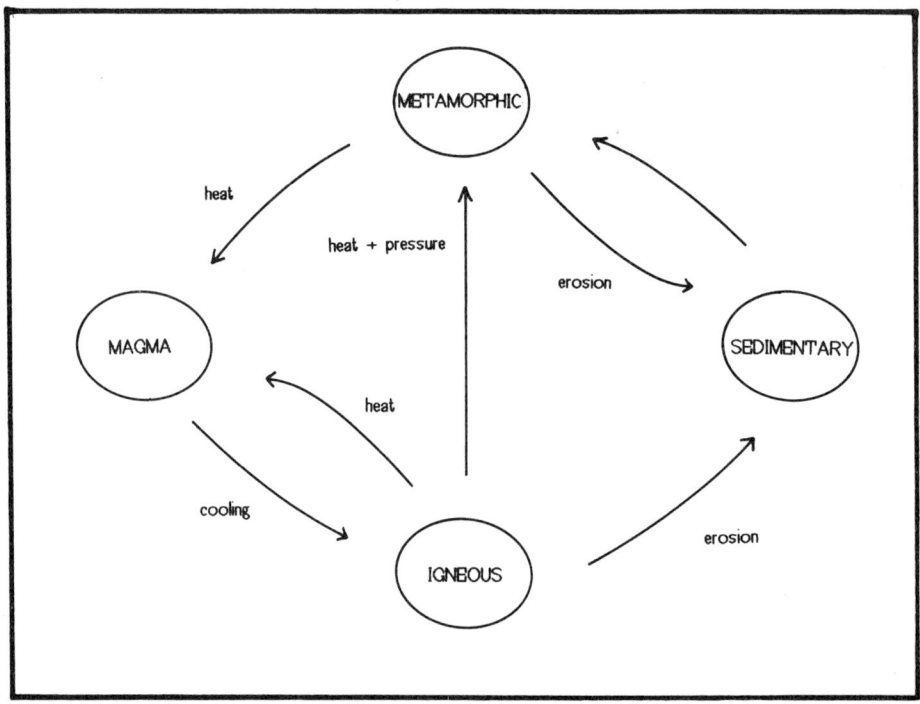

The rock cycle keeps the earth's materials in balance.

where in the rock cycle or the never-ending process that keeps the earth's rocks in balance.

Igneous rocks, from the Latin word *ignis* (or "fire"), are formed from magma from deep inside the earth. As the magma reaches the surface and cools, such as with a volcano, igneous rocks form. Basalt is a dark igneous rock formed when magma cools slowly. It is often seen on the Hawaiian Islands, where it is called pahoehoe, aa (pronounced ah-ah), or pillow lava, depending on how fast or slow the magma cools once it reaches the surface.

The second family is called sedimentary rock, named from the Latin word *sedimentum*, meaning "settling." Sedimentary rocks are made

Sedimentary rocks often form from sediment carried in mighty rivers, such as this one in Yellowstone National Park.

up of small or large pieces of rock eroded by water, wind, or ice. The fragments are eventually cemented together to form a rock of their own. This type of rock may also contain fossils, or remnants of organisms that lived long ago. Sandstone is a sedimentary rock that forms when grains of sand become cemented together. A conglomerate forms when large rounded pieces of rock are held together by a natural cement, much like manmade cement.

The last family is called metamorphic rock, from the Greek and literally translated "change of form." Metamorphic rocks are made of pre-existing igneous and sedimentary rocks that are changed by the earth's internal heat and pressure; by chemical reactions, including reactions with water; and, most importantly, by the pressure and heat generated during the formation of mountain ranges.

How do scientists know how metamorphic rocks form without seeing below the surface? They have become "scientific detectives," deducing how heat and pressure change the rock by deriving the information from various surface clues.

There are many clues. They are found in the rock layers that make up the greatest mountain ranges such as the twisted metamorphic rock of the Himalayas. Other clues about metamorphic rocks are found in large outcrops of rock around the world, such as in the highlands of Greenland and the colorful layers of the Grand Canyon. There are so many examples!

What is a Mountain?

Webster's dictionary defines a mountain as "any part of a landmass which projects conspicuously above its surroundings." But this definition may be much too simple. After all, in the lower 48 states of the United States, the highest mountain is Mount Whitney in California at 14,496 feet (4,418 meters) and the lowest mountain is Florida's Iron Mountain at 330 feet (101 meters). What a difference in size!

The definition of a mountain is often a point of view. For example, a Sherpa guide in the Himalayas once told a British climber that the 11,500 foot (3,505 meter) peaks in the Himalayas had not been named because they were "just foothills"!

In the 1800s, several European geographers described a mountain as an area at least 3,000 feet (914 meters) above the local relief. If this were true, the only mountain ranges would be the Alps, Andes, Cascades, Caucasus, Himalayas, Pyrenees, Rockies, and Sierra Nevadas! But what about mountains such as the Appalachians of the eastern United States and the Urals of the Soviet Union?

There are other disagreements about the definition of a mountain. Some people prefer to separate mountains into high and low mountains. German-speaking people use the terms *hochgebirge* ("high mountains") for mountains such as the Alps and *mittelgebirge* ("middle mountains") for mountains like those of the Black Forest. The French call these two different mountain types *hautes montagnes* and *moyennes montagnes*; while, in the United States, the terms high sierras and sierras are used. But all of these mountains vary greatly in height.

Today scientists believe that there are two main criteria that describe a mountain: the type of environment that exists on the mountain and the geology of the mountain. Both are very important.

Great mountains are often divisions between two different climates. The Andes Mountains in South America and the Cascade Range in the northwestern United States separate different environmental areas east and west of the mountains: one side is rainy, and the other side is dry.

The environment on a mountain is extremely variable. As hikers climb a mountain, they notice changes in the various types of flowers, bushes, and trees. Climbing a relatively short distance uphill, they may note that the plants are replaced by other types of vegetation. They may also observe that there are certain types of wildlife that occupy the various elevations of the mountain.

As one travels up a mountain, the change in vegetation and wildlife is often dramatic. Some plants and animals can survive the harsh winds and cold of the upper mountain regions. Others can survive only the warmer conditions at the foot of a mountain. The term *alpine* comes from the Alps and means a cold and windy zone of the forests. But along the steep slopes, sparse vegetation still grows, and animals such as mountain goats find it easy to climb and live along the rocky slopes.

Geology also helps to describe a mountain. The rocks of a mountain often tell the tale of its origin. Cracked or folded rock may be evidence of millions of years of pushing and pulling of the rock into craggy contortions much like the folded Appalachian Mountains along the eastern United States. Other mountains grow from the activity of erupting volcanoes, such as the tall mountains of Hawaii and Mt. Fuji in Japan. Mountains also originate from the faulting and uplift of rock, usually in association with the earth's restless crust. The Teton Range in Wyoming and the Front Range in Colorado are examples of faulting and uplift, respectively.

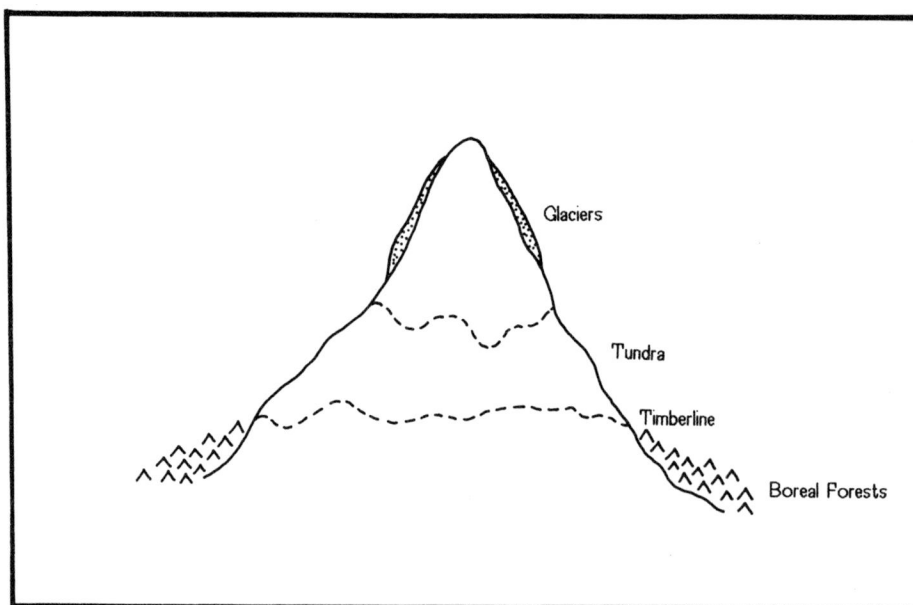

The typical mountain has many divisions as the height increases.

Some mountains may not be noticeable. Many times they are older mountains that have been worn down by erosion. This is a natural process whereby wind, ice, and water wear down the rock. The Canadian Shield, a large expanse of metamorphic rock around northeastern Canada, was once a giant mountain range. Erosion, especially by the great ice sheets of the Ice Ages, has worn down the once great peaks for millions of years!

If there were no mountains, there would be no special mountain plants or animals. There would be no mountain glaciers and no grand vistas. There would be few interesting metamorphic rocks, such as those of the Scottish Highlands or the thick beds of marble found in the rugged areas of Vermont.

There are several other reasons why these metamorphic rocks formed, the two most important being heat and pressure. How can hot rocks and pressure be responsible for such major formations of rocks and mountains? The answers are found deep inside the earth.

2

Clues in Hot Rocks

On January 23, 1973, the town of Vestmannaeyjar shook three times. This small town on the island of Heimaey in the Vestmann archipelago south of the mainland of Iceland was in for a shock: a wide fissure in the ground was sending fountains of hot rock high into the air! Everyone escaped the spray of molten rock. Within six months, the dust, lava, and ash had created a new volcano: the Eldfeld, or "Mountain of Fire."

This event was nothing new to the people of Vestmannaeyjar. Ten years earlier, a great volcanic island called Surtsey was born in the sea nearby.

The eruptions of volcanoes are evidence of the heat within the earth. There have been other occurrences. In 1943, as Dionisio Polido plowed his field in Mexico, the ground suddenly broke open with a roar. The next day the hot rock from deep inside the earth reached more than 200 feet high!

The heat within the earth is not like the heat felt on a warm, sunny day. Nor is it like the heat from a fire burning off the winter's cold. The heat deep below the earth's surface is thousands of times hotter—hot enough to melt and change rock.

Ancient peoples did not know much about the earth's interior heat.

Many believed that the earth was hollow. Others believed that it was filled with strange demons. By the 1700s, science had greatly advanced, and scientists began to look at the surface rocks to learn about the rocks in the earth's interior.

The first information about heat just below the earth's surface came from metal and coal mines. Several scientists noticed that as one went deeper into an underground mine, the temperature increased. By 1870, measurements of temperature at greater depths were made by lowering thermometers into holes in the rock at the deeper mines. The temperature change appeared to be 1°C for every 98 feet (30 meters) of descent. If these measurements were true, at a depth of 311 miles (500 kilometers) the temperature would be more than 10,000°C (18,032°F)! This finding could not be true because going any deeper would mean reaching temperatures equal to those thought to occur on the sun!

More recently, scientists have decided that the temperatures in the interior of the earth do not continue to dramatically increase as they approach the center. They have taken their data from gas and petroleum wells, and coal and metal mines. Some mines extend as far into the earth as 10,000 feet (3,048 meters). Petroleum and gas wells have been drilled to depths of 16,000 to 23,000 feet (4,877 to 7,010 meters). In 1984, one of the deepest petroleum wells was dug: the Bertha Rogers well in Oklahoma, 33,000 feet (10,058 meters) or more than 6 miles (9.7 kilometers) deep. Drillers stopped because the drill reached a molten mineral called sulfur.

Scientists have only drilled into 0.15 percent of the surface of the planet. One of the deepest holes ever drilled was in the U.S.S.R. on the Kola Peninsula. In 1984, the drill was more than 39,000 feet (11,887 meters) deep. At the 15,000 foot (4,572 meter) level, the drill passed a zone where there were vast areas of hot, flowing water, proof that the interior of the earth is a haven of hotness. Based on the temperatures registered at the many drill sights and other tests, scientists believe that the interior of the earth reaches close to 4,000°C (7,232°F).

A Peek Inside the Earth

Today scientists have determined a great deal about the hot rocks below the surface of the earth. They were aided by one great theory, an idea proposed by Sir Isaac Newton, a famous eighteenth century scientist. Newton believed that there was a force that kept objects on the earth. This is called gravity, and it has a great affect on what happens inside the earth.

Why is gravity important? Scientists believe that when the earth began to form from the early solar nebula around six billion years ago, it was very hot. As it cooled, gravity caused very dense (or heavier) material to fall to the center of the forming planet, forming the central core. Next, less dense material formed a layer around the core called the mantle. Around 4.6 billion years ago, the earth's crust formed. Today, scientists know the earth is divided

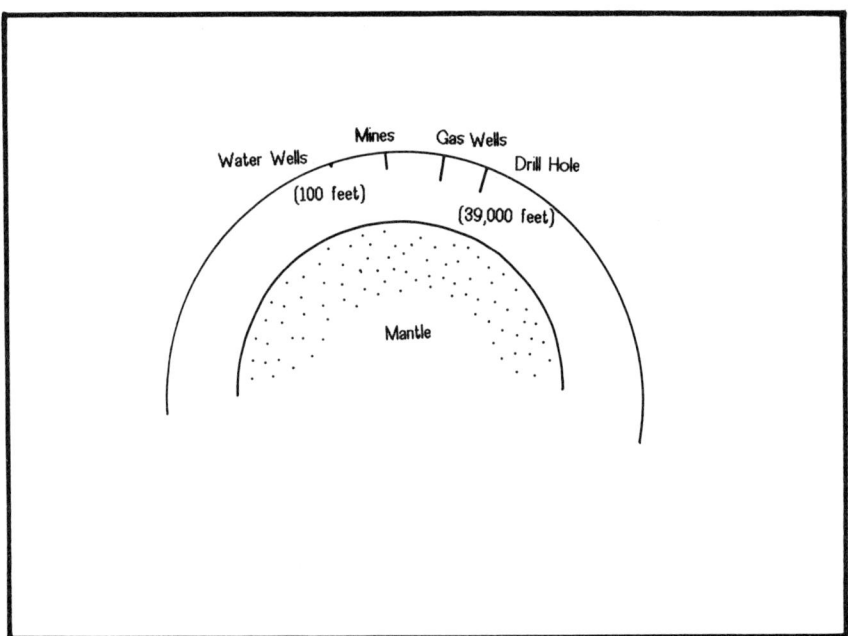

Wells and drill holes have cut into only 0.15 percent of the surface of the earth.

into these three sections that look like the layers of an onion skin.

The thickness of the earth's layers vary greatly. The core has a radius of 2,150 miles (3,460 kilometers). There are two parts to the core, the inner core, with a radius of around 750 miles (1,207 kilometers), and an outer core about 1,400 miles (2,253 kilometers) thick. Both cores are made of the metals iron and nickel. The inner core is solid, whereas the outer core is molten and moving. Scientists believe that the presence of magnetic metals and their movement may be responsible for the earth's magnetic field.

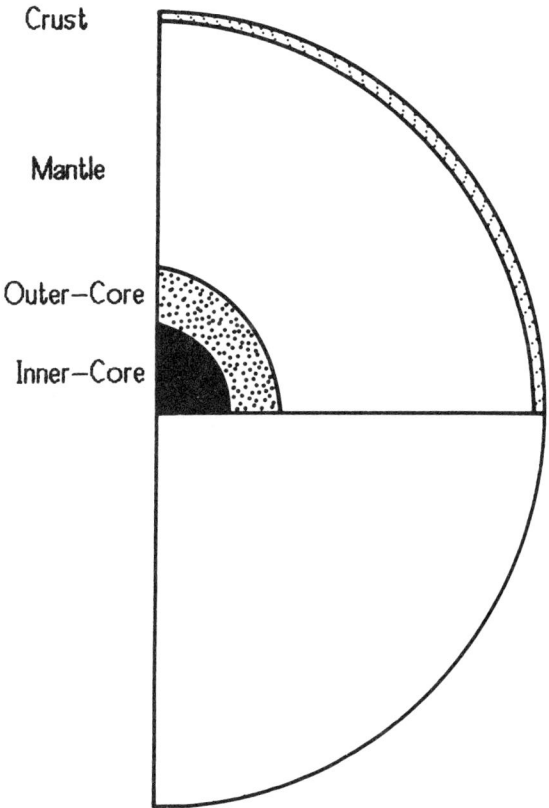

The earth is divided into layers—like the layers of an onion skin!

The core is the hottest part of the earth's interior. Temperatures range from 2,000 to 4,000°C (3,632 to 7,232°F). The pressures which are immense, are measured by atmospheres, where one atmosphere is the pressure on an object lying on the surface of the earth at sea level. In the deepest oceans, measurements are close to 1,000 atmospheres. The core pressures are close to one to three million atmospheres!

The mantle is the next layer of the earth's interior. The bottom of the mantle starts at the top of the outer core, 2,150 miles (3,460 kilometers) from the center of the earth. The top of the mantle averages around 3,950 miles (6,356 kilometers) from the center of the earth, about 1,800 miles (2,896 kilometers) thick.

Temperatures of the mantle vary from 1,000°C (1,832°F) to 2,000°C (3,632°F). Pressures in the mantle are from 10,000 to one million atmospheres. It is rich in iron and magnesium; and just above the outer core, it is very viscous, or thick and slow moving. The last 62 miles (100 kilometers) of the mantle are much more fluid.

The upper part of the mantle is in constant motion because of the intense heat generated in the earth's interior. The best comparison is a boiling pot of water in a clear glass container. As the water heats up, the cooler, heavier water falls to the bottom, while the lighter hot water rises to the top. This movement is called convection.

The earth's upper mantle responds in the same way. As the hot mantle material closer to the core heats up, the molten rock rises and cools. As the material cools, it falls back toward the core, a convection cycle that has continued inside the earth's mantle for millions of years.

The last layer of the earth is called the crust. It is the thinnest layer, only 3 to 5 miles (5 to 8 kilometers) thick below the oceans and 15 to 30 miles (24 to 48 kilometers) thick below the continents. The top layer of the crust is where humans and animals live and is the easiest layer to study.

No one can see below the surface of the earth. So how did scientists discover the layers within the planet? They used information from a field called seismology, which is the study of how the earth responds to the shakes and quakes of earthquakes.

When an earthquake occurs, natural vibrations, or waves from the quake, travel through the earth. Ancient peoples tried to explain these shakes and quakes through many legends. Aristotle, the famous Greek scientist, believed that earthquakes were caused by winds trapped inside caves. The Indians believed that the earth was on the back of a tortoise. With each step the animal took, the earth would quake in response.

Scientists know that earthquakes are caused by the restless movement of the earth's crust. When an earthquake occurs, the shaking sends earthquake waves, also called seismic waves, through the many

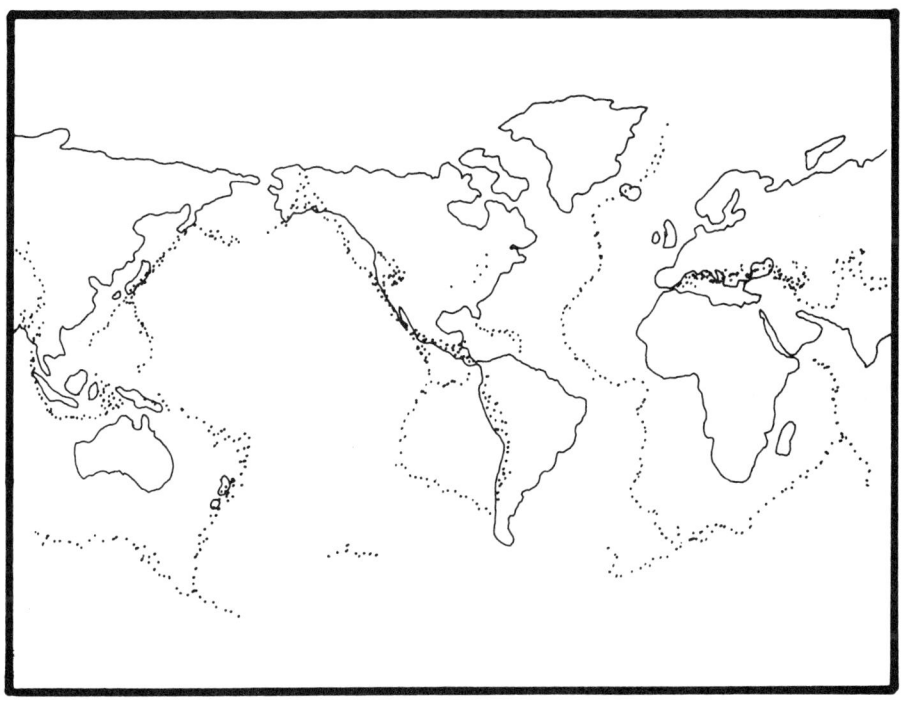

Every year most earthquakes occur in these active areas.

layers of the earth. These seismic waves can be recorded by a special instrument called a seismograph.

Scientists also use a type of seismic computer tomograph (CT) to determine the mysteries of the earth's interior. The CT is much like a scan of the human body. With a human CAT (computerized axial tomography) scan, X-rays penetrate the person from all directions. Because the bones and organs of the body have varying thicknesses, the X-rays are absorbed in different ways. The resulting image looks like a slice of the body.

By measuring seismic activity, scientists quickly found out that the earth is composed of layers of varying thicknesses. Unusually hot rock slows down seismic waves, whereas cooler rock speeds up the waves. The seismic tomographs, each stationed in various regions on the earth, act like taking a human CAT scan, giving the scientists an idea the earth's interior. Seismic waves are bent, reflected, or not transmitted at all through the earth. Because scientists know how the waves travel through certain types of rock, the graph's information can often reveal the the types of various rock layers.

The seismic information also reveals that the mantle is not a constant temperature. At various places in the mantle, there are hot and cold blobs of material. Scientists believe that these hot and cold areas help to explain plate tectonics, the theory that the earth's crust is broken into pieces or plates that move slowly around the globe, and may be responsible for the variations in the earth's gravity field.

Scientists have also found a major boundary below the earth's surface: the Mohorovicic (or Moho for short) discontinuity. Andrija Mohorovicic, a Yugoslavian scientist, discovered the division in 1909 while studying Balkan earthquake waves. He found that the material within the earth's mantle is denser than the crust. The Moho is a zone of different densities that separates the thick mantle from the thin crust. Under the oceans the Moho is not very deep, only 3 to 5 miles (5 to 8 kilometers) below the surface. The Moho is much deeper under the continents, measuring 15 to 30 miles (24 to 48 kilometers) below the surface.

The Hot Interior

By studying seismic information, scientists have determined that each layer of the earth has its own distinct properties and temperature. The heat within the earth is not constant. The rate at which temperatures change with depth is called the geothermal gradient.

What causes this geothermal gradient? Where does the heat within the earth originate or the natural heat that also contributes to mountain building and metamorphic rock? Scientists know that the heat is not from the sun or, as ancient peoples believed, from a great fire within the earth. The answer lies with something called radioactive decay, a natural process that occurs within the earth.

Radioactivity was discovered a very short time ago. In 1896, Antoine Henri Becquerel, a French physicist, discovered radioactivity by accidentally exposing several photographic plates to a rock containing radioactive uranium. Since then, more than forty radioactive materials have been discovered.

But what does radioactivity have to do with producing heat within the earth's interior? To answer that question, it is first necessary to understand radioactive decay.

Everything on earth is made up of molecules. These are tiny particles so small that they cannot be seen with the naked eye. Each molecule is further broken down into atoms (or elements), such as oxygen (O), nitrogen (N), or carbon (C). For example, water is a molecule broken down into two atoms of hydrogen and one atom of oxygen, or H_2O. An atom can be further broken down into protons, neutrons, and electrons. Protons and neutrons make up the center, or nucleus, of an atom, while the electrons circle around the nucleus. Each atom has a certain number of neutrons and protons in its nucleus.

Radioactive atoms also have a certain number of protons and neutrons. For example, in the radioactive form of the atom uranium, there are 146 neutrons and 92 protons. The total of the neutrons and protons is 238. Scientists label this radioactive element Uranium-238 (U-238). But the number of neutrons in an uranium nucleus can vary.

Thus, the number of protons and neutrons for a different radioactive uranium atom may add up to 235 (U-235). These different varieties of the same element are called isotopes.

Each isotope has a half-life or a given amount of time for half of the radioactive atoms in a rock to disintegrate. The heat within the earth results from the breakdown of all the radioactive isotopes, from radioactive uranium to potassium. For example, Uranium-238, a very abundant radioactive isotope, gives off much of the heat within the earth's mantle and core.

As the isotope breaks down, it turns into another, less radioactive isotope. Then that isotope breaks down into another isotope, and so on, until there is almost no radioactivity left in the rock. This process may take a few microseconds or billions of years! And each time an isotope breaks down, a certain amount of heat is released. There are more than one hundred elements known on the earth and many are

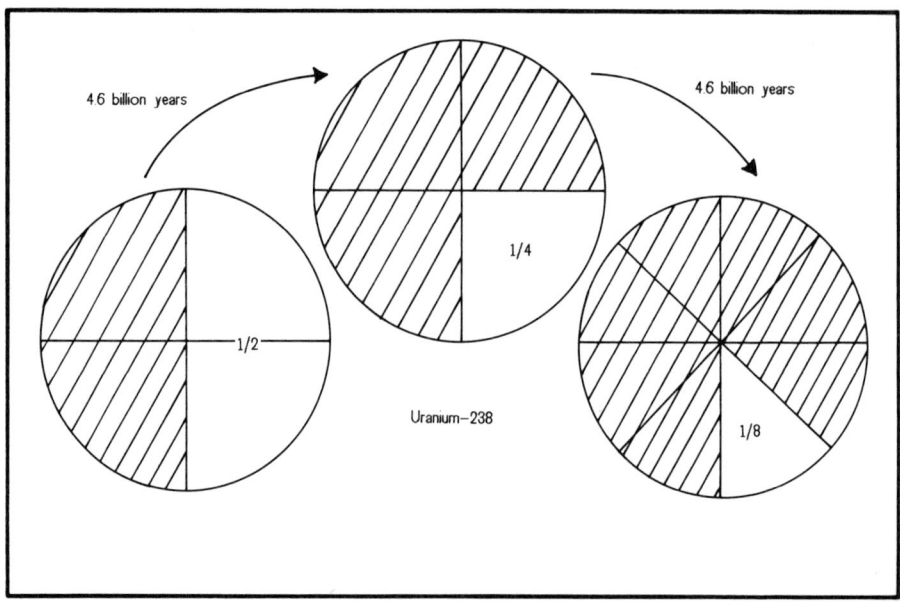

The decay of radioactive uranium over time.

radioactive. The radioactive elements within rocks and minerals help contribute to the earth's hot mantle.

Heat definitely plays an important part in the formation of metamorphic rocks and mountains, and it is known where most of the natural heat originates. But what about the next important clue in the formation of mountains and metamorphic rocks: the pressure put on rocks?

3

Press and Stress

It took many years but eventually a small piece of land moved closer to a larger landmass. Thousands of years passed, and a collision was imminent. After a thousand more years, the two chunks of land met.

This meeting of two lands was not like the collision of a hammer and nail. There was no real bang or crash. The movement of the two landmasses was too slow!

What did happen was even more amazing: The force of the contact pushed up a great mountain chain. Tall walls of rock folded into strange contortions. Other rocks slowly cracked and fractured. Though it was a very slow process, the land eventually changed from relatively flat to mountainous all within a few hundred thousand years.

Today there is evidence of this collision: the great Himalayan Mountains. The mountain chain was created when a piece of land we call India collided with the massive Asian continent. Several other mountain chains around the world formed in the same way.

Besides the building of great mountains, one of the most fascinating results of the movement of huge chunks of land is the formation of metamorphic rock from the heat and pressure of the collision. Though metamorphic rocks form in other ways, one of the most impressive is by the great pressure resulting from mountain building.

The pressure and stresses within the earth are very extreme. Pressure can result from the pushing and pulling of the earth's crust. Pressure can also be produced by the weight of layer upon layer of rock, or by underground water.

One of the most obvious ways in which rock is put under pressure is by the movement of the earth's crust. (Remember, the crust is the top layer of the earth, ranging in thickness from 3 to 5 miles [5 to 8 kilometers] below the oceans and 15 to 30 miles [24 to 48 kilometers] below the continents.) The earth is one of the most restless planets in the solar system. Its surface is constantly undergoing change. New crust is always being formed and older crust deformed. It does not take a few days, weeks, or years for the surface to move. It takes thousands and thousands of years!

Movement of the Crust

The idea that the earth's crust moves was suggested at various times throughout history. Various scientists realized that the continents appeared to fit together, just like a gigantic jigsaw puzzle. They thought that the continents may have once been pieced together. But there was little evidence to support the idea. After all, no one could explain how the continents moved.

In the early 1900s, Alfred Wegener, a German geophysicist, revived the idea that the continents move across the earth. This was called the theory of continental drift. Wegener believed that the continents were once part of a single supercontinent called Pangaea (Greek for "all lands"). His theory was that the giant continent broke apart around 200 million years ago, with the individual sections eventually drifting, turning, and twisting to their present positions.

Wegener had many clues. He noticed that there were similar fossils found on the continents of South America and Africa. He noted several mountain chains lined up from one continent to the other. And there was evidence of a cold glacial ice sheet that covered much of southern

South America, Africa, and Australia, suggesting that these three continents were once joined together.

At first, Wegener's idea was popular with other scientists. And in 1937, a South African, A. L. DuToit, wrote a special book on the earth titled *Our Wandering Continents*. But scientists soon put the continental drift theory aside. There were too many unanswered questions about how the landmasses moved around the earth.

In the 1960s, there was a renewed interest in the moving continents theory. Scientists found rock and fossil evidence of the supercontinent called Pangaea, which eventually broke apart to form two smaller continents called Laurasia and Gondwanaland. From these two landmasses came our present continents. And today, scientists keep track of the seven continents that are still on their slow journey across the earth's surface.

This rock was pressed when one side of the land moved in the opposite direction from the other.

Why does the earth's surface constantly change? The reason is that the earth is graced with an irregular crust. The entire surface of the earth is broken down into large chunks of land called plates. They are not rounded like plates on a kitchen table, but are very uneven in shape.

The earth's surface is broken into around twenty of these different plates. Each one is like a great ice floe, twisting and turning in different directions and moving only inches per year. The crustal plates carry every geologic feature known on the planet, from vast plains and plateaus to giant mountain ranges. Scientists call the slow movement of these crustal pieces the theory of plate tectonics.

As the plates move in various directions, they respond in different ways. Plates often strike each other, squeezing and folding the area of contact. If the contact is made along the edge of a continent, the result

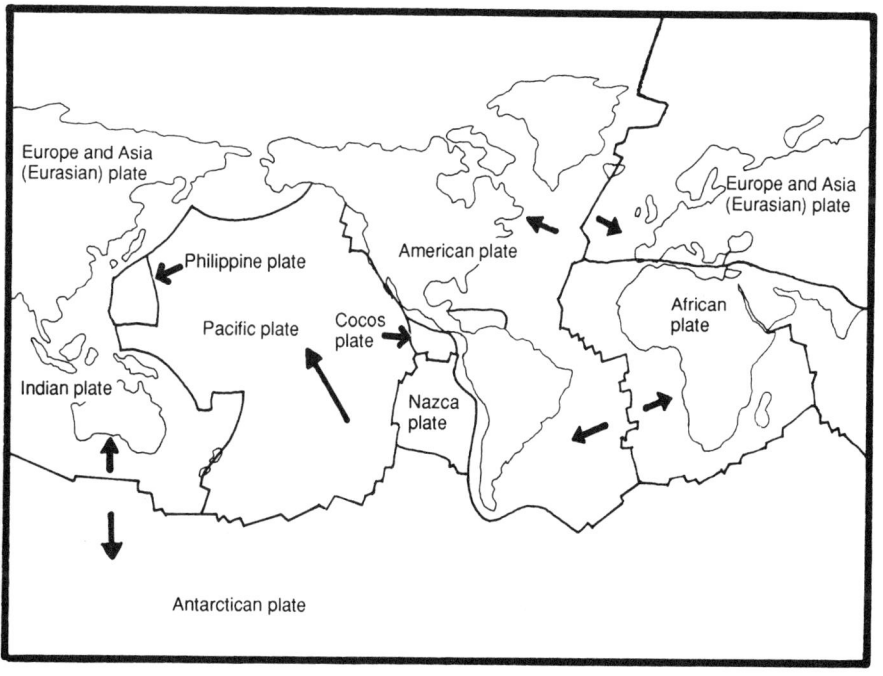

The major crustal plates of the earth.

is often the formation of great mountain ranges. Scientists call this process an orogeny. There have been several orogenies in earth's history, but they do not happen very often. The Taconic range in the northeastern United States grew rapidly during an orogeny 450 million years ago. In Europe, the Alps in Switzerland formed during the Alpine orogeny 20 million years ago, when the Eurasian plate collided with the African plate.

Wandering Plates

When plates collide, they do not always form a chain of mountains. Often one plate will slide under another plate, creating something called a subduction zone. The plate that is pushed below will eventually enter the mantle, "melting" because of the intense heat and pressure.

Usually a subduction zone forms when two plates come into contact with each other and one plate tries to undermine the other. An example of a subduction zone is the Mariana Trench in the western Pacific Ocean. This is one of the deepest oceanic trenches known on the earth, close to 6.8 miles (11 kilometers) deep.

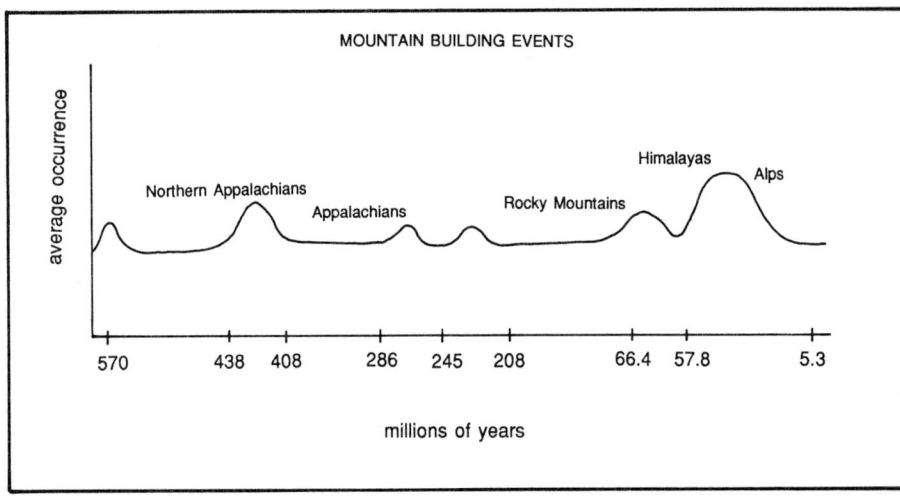

Major mountain building episodes.

In other areas, two plates may slide against each other. For example, along the California coast, there is a long, linear split in the land called the San Andreas fault. This fault, or crack in the land, is where the North American plate is moving northwest and the Pacific plate is moving southeast. The plates move very slowly. But when they do move, an earthquake may be the result.

Other plates are moving away from one another. Scientists refer to this movement as seafloor spreading. This idea was suggested in 1960 by H. H. Hess, an American geologist. Hess noted that midocean ridges of tall underwater mountains stretched through most of the major oceans of the world. Hess theorized that the seafloor was actually spreading apart at these ridges, moving sideways away from the midocean ridges.

These ridge systems deep under the ocean are essentially plates that are growing. The midocean ridges are really volcanic mountains that are continuously active. These ridges send out molten rock, constantly creating new ocean crust. The most famous growing ridge system is the Mid-Atlantic Ridge, which extends from near the Arctic Ocean, down through the Atlantic, and on to Antarctica. Iceland, a land of glaciers and volcanic activity, is right on the Mid-Atlantic Ridge, evidence that seafloor spreading is a very active process.

Why do the plates move? To answer that question, scientists first had to deduce what goes on inside the earth. When talking about plate tectonics, scientists often divide the earth into several parts: the lithosphere, which includes the crust and some of the upper mantle; the asthenosphere, partially molten material hot enough to be capable of internal "flow"; and the solid upper mantle. The rigid lithosphere plates move about on the fluid asthenosphere, like a boat on the ocean.

The movement of the lithosphere's plates is important to the formation of mountains and metamorphic rocks. But the movement of the crust is only one way in which the earth produces pressure to form metamorphic rocks. Depth is also responsible. As depth increases, so does pressure. Pressure essentially squeezes the material, giving metamorphic rocks their characteristic banding or contorted look.

What is Pressure?

Pressure is easy to understand. Try lying under several layers of blankets. The more blankets, the more pressure there is on the body. Or set a feather on a pillow. Then set a book on a pillow. The feather hardly makes a dent in the pillow's surface, but the book puts pressure on the pillow, causing it to flatten.

There are other types of pressure. One is the "pressure" noticed on one's eardrums when traveling in an airplane or diving to the bottom of a swimming pool. At any depth (or height, for that matter) there is constant pressure from all directions, a pressure that changes with greater depth or height.

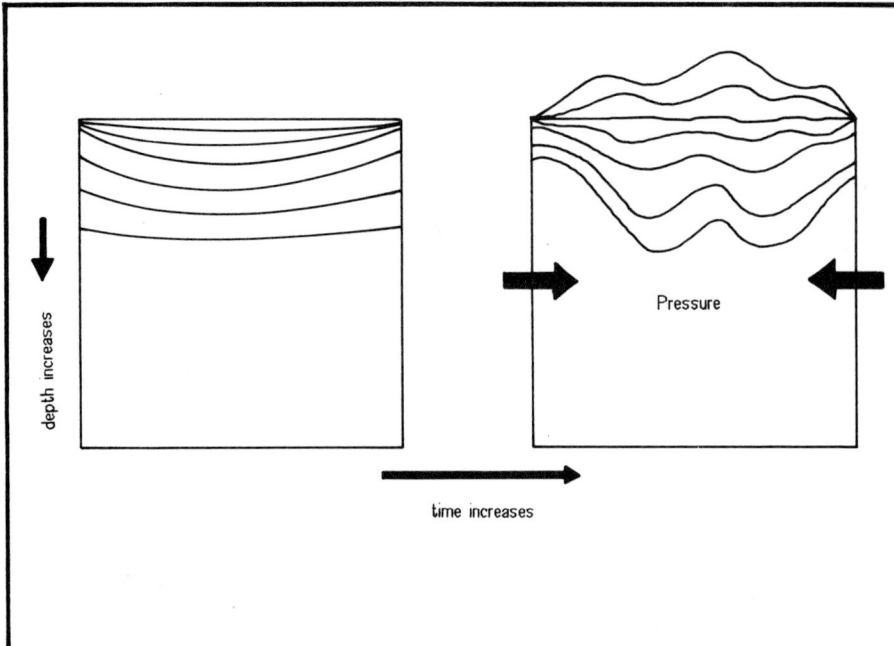

The stress of pushing on layers of rock often results in contorted mountains and metamorphic rocks. Pressure increases with depth.

Scientists measure pressure using kilobars (kbars). The normal pressure on the surface of the earth at sea level is 0.1 kilobars. If a person were able to dig a very deep hole in the earth's surface, a definite change in pressure would be noticed. The relationship between the depth and pressure is proportional. For example, at a depth of 1.2 miles (2 kilometers), the pressure is around 0.5 kilobars, while at a depth of 2.4 miles (4 kilometers), the pressure is twice that amount, at one kilobar. Having such an increase in pressure with depth puts quite a strain on rocks far below the surface of the earth.

Another pressure process is called isostasy. Though isostasy rarely builds a mountain, it can contribute to a mountain's height. For example, after the Ice Ages, the Adirondack Mountains of New York

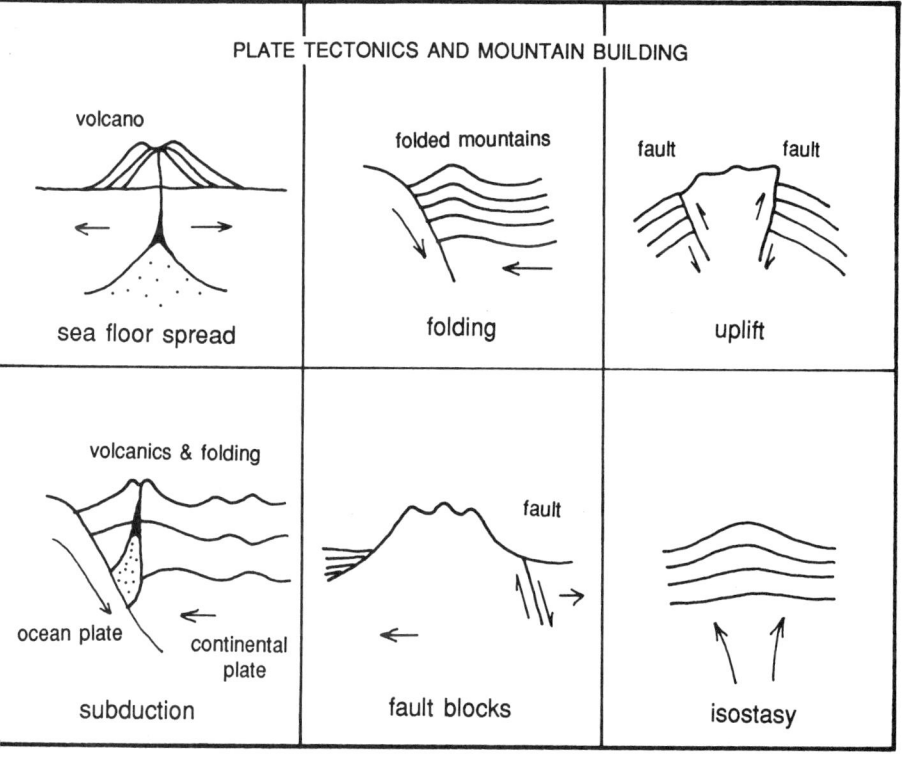

How to build a mountain.

were no longer covered by a thick ice sheet. Around 11,000 years ago, the ice sheet had retreated north. Now that the load of the ice is gone, the peaks are moving upward just less than an inch per year, rebounding from the weight of the ice sheet.

Fluids also cause pressure that contributes to the formation of metamorphic rocks. How can fluid create pressure? Take water, for instance. Water moves through the spaces (or pores) within rock. Notice how rocks along some roads seem to have spaces between them. These rock spaces are also underground where water and other fluids travel.

Fluids also travel between grains in the rock. For example, sandstone is made of small grains of sand. Each grain is rounded, making it hard for the grains to lie flat against each other.

To see how this process works, take several baseballs and pretend they are grains of sand. Gather the balls together. Notice that they touch only in one spot and they do not lie flat against each other. The spaces between the baseballs are similar to the spaces between the grains of sand in sandstone. It is no wonder that water travels right through sandstone.

Now take an area where the earth's plates collide. This massive amount of pressure deforms the rocks. And if the rocks contain fluids, the liquids are heated and expand. These fluids begin to create their own pressures, adding to the already extreme pressures of depth and compression.

4

Gneiss and Other Rocks

The peaks stand tall above the Greenland ice cap. They lie 93 miles (150 kilometers) northeast of Godthaab, Greenland's capital, and are some of the oldest mountains on the earth. They are metamorphic rocks that have been crushed, folded, and battered for the last 3.8 billion years. Scientists who study these older metamorphic rocks are hoping to learn about the early earth. After all, the earth's crust is around 4.6 billion years old, just over one billion years older than the Greenland rocks!

How do these strange metamorphic rocks form? They are the direct descendants of igneous and sedimentary rock that have been changed over time by excessive heat and pressure. Years and years later, these igneous and sedimentary rocks are often buried by sediment. As they are buried deeper and deeper, the rocks are often exposed to heat and pressure and are eventually changed; and if the rocks are exposed to the grinding forces of mountain building, they are often altered. It seems like magic. But the change of rock into metamorphic rock is actually a normal process that happens in many areas around the earth.

How Metamorphic Rocks Form

It is known that metamorphic rocks form from heat and pressure. And

Many of the rocks of the Grand Canyon have been metamorphosed by the tremendous pressures of overlying rock.

under certain situations, there can be a mixture of conditions. For instance, a metamorphic rock can form from high temperature and low pressure. Scientists refer to this process as contact metamorphism. Contact with hot magma bakes the rock, with high heat often turning the surrounding rock into metamorphic rock. For example, limestone from around Mt. Vesuvius altered because of the hot magma from the volcano. The resulting metamorphosed rock shows strange honeycombed cavities filled with minerals such as spinel, garnet, and vesuvianite. This rock was definitely changed by contact metamorphism.

Metamorphic rocks can also form in an area with high temperature and high pressure such as with mountain building. Scientists call this regional metamorphism. The mountain range known as the Alps contain numerous examples of regional metamorphism, where high heat and pressure altered the surrounding rocks. Schist (pronounced "shist") and gneiss (pronounced "nice") are common rocks resulting from regional metamorphism.

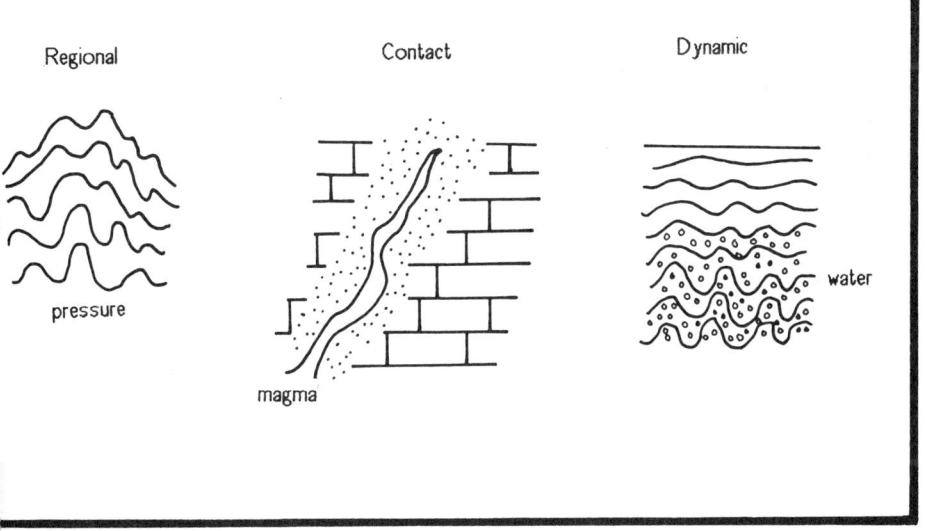

Metamorphic rocks can form in three ways: by regional, contact, or dynamic metamorphism.

The last case in which rocks turn to metamorphic rock is under variable pressures and temperatures. This process is called dynamic metamorphism. This type of metamorphism usually occurs in a small area, where the rocks are under pressure from underground water.

The search for clues to the origins of metamorphic rocks started in the late eighteenth century. There were two main groups of scientists that had theories on how the earth's internal rocks formed. One group was lead by the famous scientist and teacher, Abraham Gottlob Werner (1750-1817). His group, the Neptunists, believed that all rocks formed in water.

The members of the Vulcanists (or Plutonists) followed James Hutton (1726-1797), the famous scientist who believed that volcanic rocks originally came from a liquid rock called magma. To solve the problem, Sir James Hall, an English geologist, decided to show how rock formed deep in the earth's interior.

Hall's experiment was very dangerous. First, Hall ground down pieces of marble into small particles. Then he put the ground marble in a plugged gun barrel. This way, he could simulate heat and pressure, two factors that were present in the earth's interior. When he "fired" the marble, the pieces behaved like a liquid. His experiment proved that the Vulcanists were probably right: the rock in the earth's interior is liquid and kept fluid by heat and pressure.

Looking at Metamorphic Rocks

All metamorphic rocks, the majority formed from igneous and sedimentary rocks, are composed of minerals. Some minerals are similar to the minerals found in the original rock, while other minerals have melted and reformed, or recrystallized, into another type of mineral.

What do metamorphic rocks look like? They are very distinctive. They are strong and hard and most of their minerals are made of an element called silica.

The look of a metamorphic rock is its most tell-tale feature. Many are arranged like sheets of paper, thin layer upon layer of rock, while others are often banded or striped. The banding is due to the different colors of the various minerals within the rock.

Many times the bands look like wavy lines and are evidence of the great pressures that have crushed the rock. This type of metamorphic rock is called foliated. Foliated rocks look as if they are made of folded and bent layers, often formed by the movement of the rock as mountains form.

Metamorphic rocks can also be nonfoliated. Fossils are often found within some nonfoliated metamorphic rocks, with the fossils crushed, bent, or stretched because of the pressure. Nonfoliated rocks often look homogeneous with no real change in texture. These are often called "baked" rocks because they frequently form under conditions similar to those in which clay pottery is baked in a kiln.

The banded piece of gneiss in the upper left corner is made of the minerals (clockwise) feldspar, mica, and quartz.

How do scientists know the composition of metamorphic rocks? First, the scientist collects the rock from the field. Then it is taken to a laboratory and is split and carved into a small chunk. The next step is to polish the rock until it becomes very thin. The paper-thin almost transparent piece of rock is then mounted on a glass slide. The scientist looks at the very thin slice of rock with a special instrument called a polarizing microscope.

Another way to determine the composition of a metamorphic rock is to grind the rock into powder and look at the rock dust with a special instrument called a spectrometer. This instrument analyzes each element in the rock. After that process is completed, the rock can be identified.

There are many types of minerals in metamorphic rocks. Because these rocks are derived from other rocks, it is natural that there are

Mica can often turn to mica schist when heat and pressure are applied. The circular spots on the mica schist (on the right) are gemstones of garnet.

many minerals in them common to igneous, sedimentary, and metamorphic rock. For example, quartz and muscovite mica are common to all three types of rock.

But there are several minerals common only to metamorphic rocks. For example, minerals such as epidote, cordierite, kyanite, and sillimanite, and gemstones like garnet are common in metamorphic rocks.

Types of Metamorphic Rocks

There are many common metamorphic rocks. The most well-known is called gneiss. Gneisses are extremely banded and were originally the igneous rock called granite. One special type of gneiss called augen gneiss (from the German word *auge* or "eye") looks as if giant eyes are staring out of the rock!

Schists are very common metamorphic rocks. They were originally shale, a sedimentary rock formed mainly along lakes and rivers. Schists often contain the gemstone garnet.

Marble is a metamorphic rock that was once a sedimentary rock

---

FROM WHAT TO WHAT?
CHANGING TO METAMORPHICS

Original Rock	Metamorphosed Rock
shale	slate
granite	gneiss
sandstone	quartzite
limestone	marble
slate	schist

Metamorphic rock was originally another type of rock before it was changed by heat and pressure.

called limestone. Marbles are made of calcite and often contain other impurities such as carbon or chlorite, that give marble its various colors.

The metamorphic rock called quartzite forms when the sedimentary rock sandstone is exposed to heat and pressure. Sandstone has its origins on sandy beaches along coastlines.

One of the major types of metamorphic rocks common to the earth is slate. Slate is formed from the sedimentary rock shale, a rock that was once sediment from rivers and lakes. Slate looks much like shale but is harder and splinters when it breaks. Hornfels is another metamorphic rock originating from shale. When hot rock around volcanic areas bakes the shale, the rock turns into hornfels.

Some metamorphic rocks can be altered by heat and pressure in stages. For example, when shale is exposed to the regional metamorphic process, a new rock called phyllite develops. Additional heat and pressure turns the rock into a mica schist, which often contains minerals such as garnet.

There are many more metamorphic minerals and rocks. Talc and asbestos can form from various sedimentary, metamorphic, and igneous rocks. Large crystals called porphyroblasts can form within the metamorphic rocks. A migmatite is a mixed metamorphic and igneous rock often containing schist or gneiss and granite. And a greenschist contains mica and chlorite.

5
Mountains Around The World (and Elsewhere)

The woman wandered the hills every day with her faithful dog close by her side. As she crossed or walked around the boulders strewn across her path, she marveled at the different patterns within the rocks. Kicking a small rock with her foot, she noted brightly colored crystals. Another rock looked layered while other rocks appeared to be folded and crushed. What was wrong with these strange rocks? Why did they look as though they had been twisted?

The traveler was in the great Scottish Highlands, an area composed of metamorphic rocks that are close to 500 million years old. The Highlands are not the only place where large blocks of metamorphic rocks are found. The Adirondack Mountains of New York contain them, as do several islands in the southwest Pacific Ocean, the Appalachian Mountains along the east coast of the United States, many of the mountains of Japan, and parts of southeastern Canada, to mention only a few.

Why are mountains important? Without them, there would be a great gap in history, literature, and art. There would be no story of Hannibal crossing the Alps or of Lewis and Clark bringing back stories of the great mountains today called the Rocky Mountains. James

Hilton's fictional story of Shangri-La, a mysterious city supposedly in the Himalayan Mountains, would never have been written. Author Jean Jacques Rousseau's description of the mountains of Switzerland had an almost revolutionary impact on the way Europe viewed nature in the late nineteenth century. And the Japanese culture has always regarded mountains as sacred, an idea very prevalent in Japanese art over the centuries.

For centuries, mountains have been barriers, separating countries, languages, and philosophies and creating more diversity among human cultures. The Himalayas have always been the boundary between the diverse cultures of China and India. Along the western Pyrenees between France and Spain live the Basque, a people whose origin is unknown. These people are expert shepherds of the mountains and have developed their own language and diverse culture based on their mountain heritage.

The Legends of the Mountains

Mountains have been both held in awe and feared. Many primitive peoples believed that the mountains were good, since streams from the mountains brought water to the villages. Thus, the mountains were often thought to be the source of life.

Many ancient peoples also believed that the mountains were a place where the earth met the heavens. The mountains were a source of mystery often surrounded by clouds, lightning, storms, winds, and cold. To many North American Indians, the mountaintops were the sites of the spirits.

Early treks to the higher mountain peaks were often very frightening. The ancient humans did not know about high altitude sickness. The dizziness and shortness of breath must have made the walkers believe they were unwanted and were traveling on sacred ground.

Not all mountains were considered good. Many primitive societies believed that the mountains were the home of powerful gods or devils, especially if the mountains were volcanic. If there was a major eruption, ancient humans believed that the gods were angry. Many sacrifices and special ceremonies were then conducted to appease the mighty gods of the mountain.

There are also very strange legends attached to mountains, some still in force today. One such story is that of Yeti or the Abominable Snowman of the Himalayas. Yeti is derived from the Nepalese word *yeh-teh*, or smaller ape. The name "Abominable Snowman" first made its appearance in 1930, mentioned by a journalist in Calcutta. But this hairy, tailless wildman of northern Nepal was reported as far

This medicine wheel was built high atop a mountain sacred to a North American Indian tribe.

back as the early nineteenth century when a British resident noted that some of his porters had been frightened by the sighting.

By the early 1900s, large footprints were found at the 21,000 foot (6,400 meter) level of the Nepalese mountains. And by 1925, there was a sighting of the strange creature on a glacier at 15,000 feet (4,572 meters). A yeti-like creature was also reported in 1979, at the foothills of the Himalayas in China. Another yeti-like creature is reported to live in the vast forests of the mountain regions of central China.

The legends grew as more people claimed to see the yeti and as more books were written on the subject. The sightings have become an important part of Himalayan Mountain legend and folklore. But scientists debate the actual existence of a yeti creature or creatures.

Another mountain legend originated in the mountains of western North America. Sasquatch, or Bigfoot, is said to roam the mountains. And there may be more than one creature, for there have been thousands of sightings by hunters, American Indians, trappers, and others. Some people have fuzzy photographs of Bigfoot and others have tape recordings of its vocalizations. But many scientists believe that the evidence is fake.

However, legends of Bigfoot abound. Bigfoot is considered to be taller than the average man and is covered with a thick layer of hair. Most of the reports are from the heavily forested and uninhabited regions of the Pacific Northwest, from northern California to southern British Columbia. As long ago as 1784, a newspaper reported the existence of a "huge, manlike, hair-covered" creature captured by the Indians in Manitoba. And the reports have continued ever since. There is even a Bigfoot Information Center in Oregon.

Where are the Mountains

Most of the mountain chains around the world consist of metamorphic rocks. Since close to 10 percent of the earth's surface (or 25 percent of the area of the continents) is covered with mountains, there are quite a few places to look for metamorphic rocks!

Most mountain rocks have been pushed and pulled by forces deep within the earth, with stresses that have continued throughout most of the earth's long history. A glance at the mountain chains around the world reveals that most of the highlands tend to run in long, somewhat straight belts along the margins of the continents.

The Caledonian-Appalachian belt of mountains stretches from Greenland, through northern Norway, then on to Scotland via the Shetland Islands, then south to Wales, Ireland, and the eastern seaboard of North America. This chain of mountains originated more than 400 million years ago when the continents were closer together. Many areas of the mountains are composed of metamorphic rock and are testimonies to the stress and strain the rock had undergone during mountain building.

There are also long chains of mountains located under the oceans. There are volcanic cones, broad plains, great trenches, canyons, and deep basins in the ocean. If all the water were removed from the seas, it would resemble what is seen on land. In addition, ocean mountains can be found along the ocean ridges, continuous rocky ridges that extend in many areas around the world's oceans. These ridges are most prevalent in the Atlantic, Pacific, and Indian Oceans. Many ocean mountains are the result of volcanic activity. Ocean islands, like the Hawaiian Islands, are the result of tons and tons of molten, or liquid, rock building tall volcanic mountains.

Around the Solar System

What about the rocks and mountains on the other planets and satellites of the solar system? Do other bodies have metamorphic rocks covering them? Or do the planets and satellites have vast mountain ranges like those on earth?

So far, only one other member of the solar system has been visited by humans: the moon. Between the late 1960s and the late 1970s, twelve astronauts walked on the moon. They were carried there by the Apollo spacecraft and landed on the moon using the lunar excursion module. It was a three day trip to get to the moon, and it took three days to return to earth. The six missions of men to the moon gave scientists a better understanding of the earth's closest neighbor.

While the Apollo astronauts were on the moon, they collected more than eight hundred pounds of rock from many different sights. For example, the Apollo 14 astronauts landed in a smooth area called the Fra Mauro Peninsula. The Apollo 15 landed near the very rugged terrain of the Apennine Mountains and a long winding "gully" called Hadley Rill. The Apollo 16 astronauts collected rocks from the rugged and cratered Descartes Highlands.

When all the rocks were returned to the earth, scientists discovered that most of the rocks were of igneous origin, not metamorphic. Many of the rocks were pitted, evidence that the rocks had been bombarded by meteorites and other space objects. The youngest moon rocks were less than a million years old, whereas the oldest rocks were close to 4.2 billion years old.

Unlike the earth's moving crust, the moon has no active surface. Though there are mountains and highlands on the moon, there are no plates riding on the mantle as on the earth. How do the scientists know this? Because the Apollo astronauts also left experiments on the moon to detect moonquakes. Few quakes occurred, evidence that the moon is a very quiet place.

Scientists believe that the moon is like a big fossil in space. Some rocks collected by the Apollo astronauts had lain exactly in the same position for longer than humans have existed on the earth. In fact, the moon has changed little since its formation.

It is thought that most of the moon's mountains were formed when great space objects struck the moon, sending tons of soil, rock, and dirt high above the moon. As the soil settled down, highlands were formed. Other mountains on the moon may have formed when the space objects that formed the craters pushed up soil and rock on the surface. And still other highlands were created early in the moon's history when volcanic action was an important process on the young moon.

The other planets and satellites have been explored using visual means. Many unmanned spacecraft, including the Pioneer, Viking, and Voyager spacecraft, have sent computer-generated pictures of the surfaces of various planets and satellites back to the earth. These images look very strange when compared to the earth.

Looking at the images of the other planets and satellites, scientists agree that these bodies have had much different histories from the earth. Of the planets explored, some contain great mountain chains, such as the mountainous and knobby terrain of Mars or the hilly and cratered areas of Mercury. It is believed that many of these mountains were formed from the impact of space objects on the planet's surface.

There is evidence that the earth has been hit by meteorites.

Mars also has great volcanic mountains. One volcano called Olympus Mons towers to a height of 15 miles (24 kilometers) above the Martian surface and is three times the height of Earth's tallest mountain, Mount Everest!

Scientists believe that Venus may be the only planet besides earth that has experienced plate tectonics. The radar images of the planet show large bodies of rock, much like the continents on earth. Scientists often wonder if Venus has crustal plates and if the planet has metamorphic rocks.

Some of the smaller satellites of Jupiter, Saturn, Uranus, and Neptune also show evidence of mountains formed by the movement of the satellite's crust. Miranda, a moon of Uranus, is an amazing place filled with strange chevron-shaped ridges and cracks. Triton, a moon

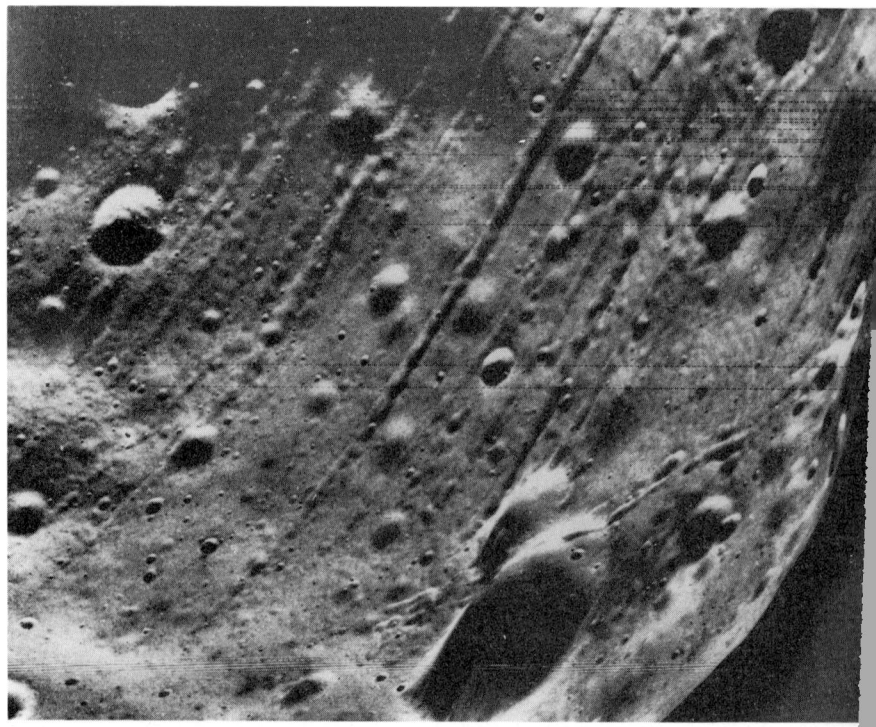

This satellite of Mars, called Phobos, probably contains few metamorphic rocks.

of Neptune, also has craggy ridges on its icy surface. Scientists believe that both crusts were at one time very active. Could there be metamorphic rocks on Miranda and Triton?

There is a great deal of evidence that shows many mountainous regions exist on the other planets and satellites of the solar system, but scientists will not know if there are metamorphic rocks on these bodies for a long time. In order to find out what types of rock exist elsewhere, unmanned and manned missions to the other bodies of the solar system must take place.

The great mountains of the world and the strange-looking rocks are reminders of the earth's constant contact with heat and pressure. But how have humans benefited from the two results of heat and pressure: the mountains and the metamorphic rocks?

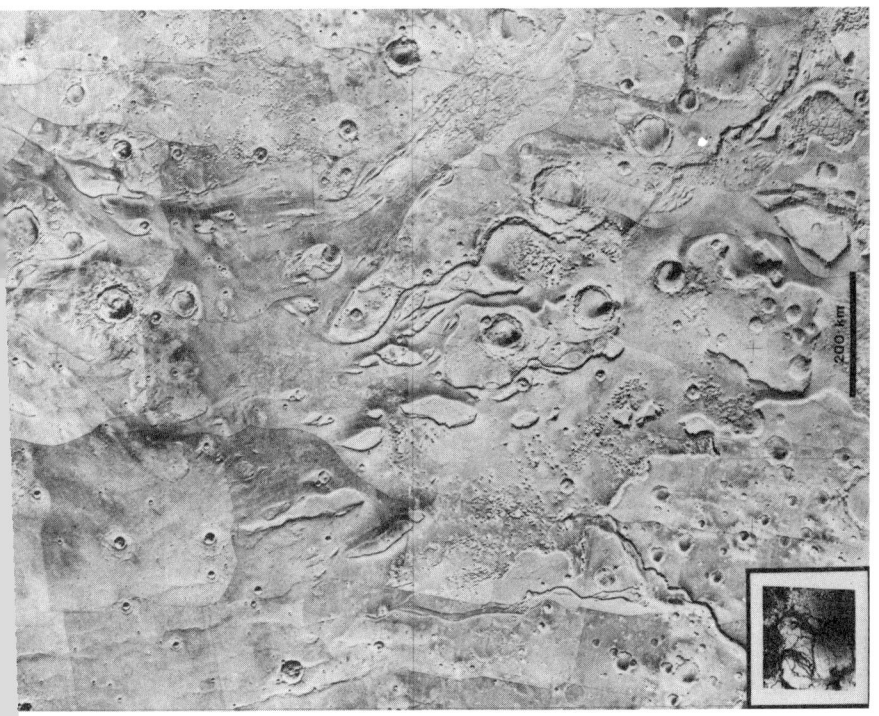

The Martian surface looks cracked and worn—does it have any metamorphic rocks?

6
Living with Metamorphic Rocks

As the miner gets to work, he notices how greatly the dark gray hill contrasts with the blue sky. His eyes wander over the area of his next task: to drive his truck up the steep gravel road, past the many rows of rock terraces, and to park just outside the underground mine. There he will pick up a load of rock and head toward the processing plant nearby. It is a normal day for this miner in North Wales as he helps to mine slate, a metamorphic rock that is used for roofing shingles, stairs for homes and businesses, tabletops, trimming stones, and chalkboards.

Humans have long mined the earth. The first earth materials used by humans were nonmetallic. Such materials as flint, quartz, chert, quartzite, talc, or limestone were used for weapons, utensils, and carvings. For example, an old mine in New York still holds traces of Native Americans, who took flakes of flint and chert and turned the minerals into spearheads and scrapers.

It is estimated that around 7000 B.C., early humans used several special minerals for jewelry and ornaments, including quartz, obsidian, jasper, amber, calcite, and amethyst. Other minerals, such as mineral paints, were used mainly for body painting during tribal rites.

Most minerals were found only in certain areas, and humans used the materials that were available where they lived. Materials were also traded among tribes. Many humans traveled long distances to trade goods, including hardened stones and minerals not readily available elsewhere.

Clay was also used, mainly for pottery and eventually for bricks. Clay figures dating back to more than 20,000 years ago have been found. Early Asiatic and African homes were made of clay and, eventually, early homes in America were built from clay bricks.

Large stones were also used to build structures such as homes and castles. On a larger scale, the pyramid of Giza contains 2,300,000 blocks of stone averaging 2½ tons apiece!

As civilization progressed, so did the desire for gemstones. They were greatly prized. Colors were the most important aspect of the jewel, including the deep blue of lapis lazuli, the purple of amethyst, the green of malachite, and the light blue of turquoise. Other gems were used for their ability to polish and shine, including agates and garnets.

The mining and use of gemstones was important to the early Egyptians, Babylonians, Assyrians, and Indians. Living or dead, important Egyptians were bedecked with jewels. Scientists have found evidence of jewels in the various Egyptian tombs that have been opened. King Tutankhamen, (fl.1350 B.C.), is one of the more famous Egyptian kings because of the great number of jewels found in his tomb.

Early Search for Metals

Scientists believe that gold was the first metal found. It has always been one of the most prized metals and is mined from deep manmade holes in the earth's surface. It can also be extracted from streams and rivers, as small pieces of the metal are carried along the channels.

Through history, there have been many methods of extracting gold from the earth. Stories are told of the country of Saones, where streams from the mountains carried gold. The barbarians, as the people were

Metals are often found in deposits such as this igneous dike.

once called, collected the water in long troughs pierced with holes and lined with the fleece of sheep. The fleece linings trapped the gold and were hung to dry in trees. When they were dry, the gold would be beaten out of the fleece. (Many historians believe that from this practice sprang the Greek legend of how Jason and the Argonauts searched for the Golden Fleece near the shores of the Black Sea.) Today similar methods are used in South America to extract gold from streams.

Another way of finding gold is by panning. This method was very popular during the California Gold Rush of 1848. Gold prospectors would use shallow pans and fill them with sediment and water from a river. The prospectors would shake the pan, hoping that the heavy gold would sink to the bottom. Today rock collectors use this method in areas such as North Carolina, Oregon, and California to find small pieces of gold.

Many other metals were mined, including copper, silver, and iron. These metals were easy to twist and bend into desired shapes, and pound into patterns. Copper was found around 18,000 B.C. and by 4,000 B.C., was widely used in Europe. Silver was once mined in southern Spain by Hannibal, the famous Carthaginian general who invaded Italy by crossing the Alps. The yield was 300 pounds of silver a day!

The Importance of Metamorphic Rocks

Why are certain stones, metals, and gemstones important to the study of metamorphic rocks? Because many of the stones used by early (and recent) humans were derived from metamorphic rocks. Today scientists know much more about extracting minerals and rocks from the earth and are also developing more efficient ways to use the raw materials.

For example, at one time slate was likely to be found in any town, especially on roofs and in sidewalks. It is durable and comes in colors ranging from gray to red and green. But today, it is too expensive to use slate as a roofing or sidewalk material. Now it is mainly crushed and used in building materials.

Graphite is another metamorphic rock that has many uses. Once mistaken for the metal lead, it is frequently called "black lead." Graphite is a soft, greasy-feeling black material made mainly of the element carbon. It is found in marble, gneiss, schist, and quartzite. Deposits are found in many places, including Alabama, Madagascar, and Sonora, Mexico. This material can mark paper and is found in pencils. The term *graphite* comes from the Greek word *graphein*, meaning "to write." It is also used as an ingredient in paints.

Asbestos is a fibrous mineral that forms under metamorphic conditions. Asbestos was once used extensively for electrical insulation,

These miners searched for metals in the rock formations of Telluride, Colorado during the late 1800s.

fire-proofing, and insulation for schools, industrial buildings, and homes. But in the early 1980s, asbestos fibers were found to cause serious lung problems in humans. Since that time, asbestos has been removed from most structures and is no longer used as insulation. One of the world's largest deposits is in Quebec, Canada. Other significant deposits are found in Zimbabwe, South Africa, and Russia.

Talc is also a material of metamorphic origin. It is very soft and is of hardness one on the Mohs' scale of hardness. Talc is widely used as a lubricating material and is the main ingredient in talcum powder. Talc deposits are found in Ontario, North Carolina, and Austria.

Another form of talc is called soapstone, which is also a soft rock, mainly composed of talc but containing other minerals such as quartz or chlorite. Sculptors often use soapstone or talc to carve intricate art forms and figures. In New England, where many deposits are found, stoves made of soapstone are common!

There are several special minerals associated with metamorphic rock. Three in particular have very strange names: andalusite, kyanite, and sillimanite. These minerals are able to withstand high temperatures and they are very resistant to shock. In addition, they are often used in ceramics and high-temperature insulation. For example, many spark plug companies use andalusite for their spark plug cores. The United States, India, and Kenya are the major producers of these minerals.

Metals and Gemstones

There are few metals associated with metamorphic rocks, as most of them are common to igneous rocks. But gold is one exception. Since 1879, the Homestead Mine in South Dakota has produced millions of dollars worth of gold. The gold was formed when a schist was replaced by hot igneous rock.

Gems from metamorphic rocks are very desirable. Garnets are the most prevalent and are used both in jewelry and in industry. Garnets are usually found in gneisses and schists, with colors ranging from yellow to deep red.

One of the largest deposits of garnets in the world was located at Gore Mountain in upstate New York. At one time, 90 percent of the world's industrial garnets came from there. The mine has yielded giant garnet crystals up to two feet in diameter! The mining operation was recently closed and moved to nearby Ruby Mountain, another area filled with garnets. The garnets around this area of the Adirondack Mountains are usually crushed and used as garnet paper for woodworking and glass polishing.

The gemstone lapis lazuli, once prized by ancient civilizations, is also found in association with metamorphic rocks. Ancients used the stone for mosaics and beads. Today the deep blue stone is usually used to adorn rings, necklaces, and bracelets. Most of the stones come from Afghanistan.

The gemstone ruby is also found in metamorphic rocks. Ruby is often faceted with delicate cuts in the stone, giving the rock a glittered appearance. Rubies are very hard, and they usually are very small. They are often red and are used for jewelry and for grinding and polishing in industry.

Rubies are found in Sri Lanka, where they have been extracted for more than 2,500 years. But the finest rubies come from the country of Burma. Rubies are not mined, but are extracted from the sediment around a former lake area. Burmese kings worked these deposits for centuries before larger operations began in 1889. Other precious stones, including sapphires and tourmaline, are found with the Burmese rubies.

Emeralds, too, are often found in association with metamorphic rocks. They are considered one of the most valuable of all precious stones and are usually a deep green color.

One of the largest emeralds was found in North Carolina. In the Ural Mountains, emeralds in schists are one of Russia's most important gemstones. Emeralds are also obtained from mica schists in Austria. And in Egypt, near the Red Sea, emerald mines were worked some 2,000 years ago, with the mine shafts sunk in a mica schist.

Mountains are important to everyone on earth.

There are other less known gemstones associated with metamorphic rocks. Spinel is a reddish color, considered by some one of the most beautiful gems in the world. Spinels are mainly from Ceylon and Afghanistan. One spinel gem, set in the English crown, is said to have been given in 1367 to the Black Prince, or Prince Edward of Aquitaine, the black-armored prince who fought in the battle of Crecy in 1346 when he was only sixteen years old.

Metamorphic rocks have been an important part of the earth's history and makeup. But mountains are also important. They have given the earth different environments. They have contributed greatly to human history and they remind everyone of the vast differences for which the earth is famous.

Study the mountains and metamorphic rocks of the world. They are as diverse as the birds, animals, trees, and flowers of nature. And they are just as necessary. Together mountains, metamorphic rocks, and humans have lived in harmony and, hopefully, they will continue to do so in the future.

Glossary

asthenosphere—A layer of the earth below the lithosphere that is usually considered the upper mantle and is partially molten.

atoms—Small particles made up of distinct combinations of protons, neutrons, and electrons. Atoms make up all material on earth. For example, a molecule of water is made up of one atom of oxygen and two atoms of hydrogen.

computer tomograph—A computer generated "picture" of the insides of an object, based on thicknesses within the object.

contact metamorphism—When rock is altered to metamorphic rock by contact with hot material such as magma.

core—The central portion of the earth, divided into the inner and the outer cores.

crust—The outermost layer of the earth averaging about 22 miles (35 kilometers) in thickness.

diamond—The hardest substance on Mohs' scale of hardness, ranking number ten.

dynamic metamorphism—When rock is altered to metamorphic rock by varying temperatures and pressures.

earthquake—The sudden shaking that results when the earth's crust moves during volcanic or tectonic activity.

geothermal gradient—The rate of increase in temperature as one goes deeper toward the earth's center.

geysers—Natural hot springs from which jets of steam or water are ejected, often associated with volcanically active areas.

Gondwanaland—One of two landmasses that were once part of the supercontinent of Pangaea, with Gondwanaland located in the southern hemisphere of the earth.

gravity—The force or pull of one body on another due to their masses.

half-life—The amount of time that it takes for half the atoms of a radioactive substance to decay.

igneous rock—A rock type representing materials that are formed from molten rock (magma).

isotope—Two or more variations of an element that have the same number of protons but a different number of neutrons; often involved in radioactive decay.

Laurasia—One of two landmasses that were once the supercontinent of Pangaea, with Laurasia in the northern hemisphere of the earth.

lithosphere—The outer layer of the earth, measuring about 62 miles (100 kilometers) in thickness, and also called the solid crust and part of the upper mantle of the earth.

magma—Hot rock from inside the earth that often reaches the surface in the form of volcanic activity; igneous rock forms from magma.

mantle—The middle layer of the earth's interior between the crust and the core; this moving layer may be responsible for the movement of the earth's crust.

metamorphic rocks—Sedimentary and igneous (and often already altered metamorphic) rocks that are changed by various combinations of heat and pressure.

mineral—A naturally occurring, homogeneous compound composed of one or more chemical elements.

Mohorovicic discontinuity (Moho)—A uneven imaginary line that separates the crust from the fluid mantle layer.

Mohs' scale of hardness—A scale invented by F. Mohs with ten minerals that represent various degrees of hardness.

molecules—Smallest particle of a substance that represents one or a combination of atoms; for example, two atoms of the element hydrogen and one atom of oxygen forms a molecule of water.

mountain—A single, steep-sided feature with elevations higher than a "hill," but usually with many local definitions; a mountain is often defined in terms of its special wildlife, vegetation, and geology.

neutrons—Particles that are found in the nucleus of all atoms.

orogeny—Periods of mountain building that occur as the result of the movement of the earth's crustal plates.

Pangaea—The hypothetical supercontinent that scientists believe once contained most of the present continents.

plate tectonics—The theory that the earth's crust is broken down into plates that move like ice floes in a glacial river.

porphyroblasts—Large crystals that can form from rapid cooling in metamorphic rocks.

protons—Positively charged particles that are contained in the nucleus of all atoms.

radioactivity—The decay of radioactive elements caused by changes in the nucleus and the giving off of energy and charged particles.

regional metamorphism—Rocks altered over a large area by heat and pressure; especially associated with mountain building.

rock cycle—The cycle that keeps the earth in balance by the constant forming and destruction of rock on the surface.

rocks—Any naturally forming materials made up of two or more combinations of minerals.

sedimentary rocks—Rocks that are formed by the erosion, then the cementation, of rock.

seismology—The study of how the earth shakes and quakes; also used in connection with studies on the interior of the earth.

talc—The softest mineral on Mohs' scale of hardness.

volcano—A vent through which hot magma (or liquid rock) and steam from within the earth reach the surface.

Further Reading

Bain, Iain. *Mountains and People*. Morristown, N.J.: Silver Burdett, Ginn, Inc., 1982.

Barnes-Svarney, Patricia L. *Clocks in the Rocks: Learning About Earth's Past*. Hillside, NJ: Enslow Publishers, Inc., 1990.

Rapp, George, Jr. and Laura L. Erickson. *Earth's Chemical Clues: The Story of Geochemistry*. Hillside, N.J.: Enslow Publishers, Inc., 1990.

Ridpath, I., ed. *Minerals*. Reading, M.A.: Addison-Wesley, 1975.

Schackley, M.L. *Wildmen*. London: Thames & Hudson, 1983.

White, A.T. *All About Rocks and Minerals*. New York: Random House, 1955.

Index

A
Abominable Snowman, 43-44
Adirondack Mountains, 31-32, 41
Alps, 5, 11, 12, 28
Aristotle, 19
asthenosphere, 29
atoms, 21-22

B
Becquerel, A. H., 21
Bigfoot, 44
Black Prince, 58

C
Canadian Shield, 13
computer tomography, 20
continental drift, 25
core, 16, 17, 18
crust, 12, 16, 18, 20, 25, 27, 46

D
Dante, 6
DuToit, A.L., 26
diamond, 7

E
earthquakes, 19, 29
elements, 6, 7, 21, 23

G
gems, 55-56, 58
geothermal gradient, 21
gold, 51, 53

Gondwanaland, 26
gravity, 16
Greenland, 33

H
half-life, 22
Hall, J., 36
hardness (Mohs' scale), 7
heat
    earth's interior, 14, 18, 20, 21
    radioactivity, 21, 22
Hess, H.H., 29
Himalayan Mountains, 10, 11, 24, 42, 44
Hutton, J., 36

I
Ice Ages, 13, 31
igneous rocks, 8, 9, 55
isostasy, 31
isotopes, 22

K
Kircher, A., 6

L
Laurasia, 26
lithosphere, 29

M
mantle, 16, 18
metals, 51-53, 55
metamorphic rocks and minerals, 10, 33, 35-40

metamorphism
  contact, 35
  dynamic, 36
  regional, 35
Mid-Atlantic Ridge, 29
minerals, 8
  definition of, 6
  hardness of, 7
  types of, 35, 39, 40, 50, 55
mines, 15, 50-53, 55, 56
Mohorovicic, A., 20
Moho discontinuity, 20
Mohs, F., 7
moon, 45-46
mountains
  definition of, 10-11
  environment and geology of, 11-12
  formation of, 24, 27-28, 29, 31-32
  location of, 44-45
  on the moon, 45-46
  on other planets, 47-49

N
Neptunists, 36
Newton, Sir Isaac, 16

O
orogeny, 28

P
Pangaea, 25, 26
planets, 45, 47-49
  Mars, 47, 48
  Venus, 48

plate tectonics, 20, 27-28, 29

R
radioactivity, 21-23
rock cycle, 9
rocks, definition of, 7

S
Sasquatch, 44
satellites (of planets), 47-49
Scottish Highlands, 13, 41
seafloor spreading, 29
sedimentary rocks, 8, 9-10
seismology, 19, 20
solar system, 45-49
subduction zone, 28

V
volcanoes, 6, 9, 14
Vulcanists, 36

W
Wegener, A., 25-26
wells, 15
Werner, A.G., 36

Y
Yeti, 43-44